(Photo by Austin Post

To Caroline,
with much
appreciation for
all your fabulous
music!

Don

TABLE OF CONTENTS

PREFACE

My experience with the 1980 explosion of Mt. St. Helens was up close and personal—I was flying circles over the summit for 2½ hours just before the explosion and missed being blown out of the sky by about 20 minutes. I had planned to fly from Seattle to Mt. St. Helens at seven o'clock in the morning of May 20, but Wilbur Johnson, the pilot, loved sunrises and convinced me we should leave at five o'clock instead of seven. The summit of the volcano had been subsiding substantially and we wanted to check the subsidence of rock at the summit and photograph the vent area. We arrived at St. Helens at about 5:30 a.m. and flew tight circles a few hundred feet above the summit crater until a little after 8:00 a.m. The volcano was very quiet the whole time, somewhat unusual from our earlier flights where every time if we flew around the mountain long enough, we would be rewarded with an ash eruption. With no sign of any activity, we decided to head back to Seattle. About 20 minutes later, we felt a sort of 'whumph' and the plane jumped. We jokingly said to each other, "gee, wonder if the volcano has blown up," not knowing that that was indeed the case. We landed in Seattle and I drove to Edmonds where my wife was waiting. When I walked in the door, she said 'have you heard the news about St. Helens? The radio was broadcasting an account of a big explosion of St. Helens. I said something to the effect of "Darn, we missed it!" I called Wilbur, my flying companion,

and went back to Boeing field. We flew back to the mountain and photographed the eruption and debris flows all the rest of the day. Many of the photos in this book are from that flight.

I subsequently flew to St. Helens many times, including numerous helicopter flights. With every helicopter flight, we would fly around the mountain for half an hour or so in awe of what we saw before landing at many place to collect samples for a research project funded by the National Science Foundation. All of the ground photos in this book were taken during these helicopter trips.

Looking back 36 years later, I have to say that this was the most awesome event I have ever witnessed. I've seen volcanic eruptions in Hawaii, but this was altogether different. Cauliflower clouds of ash would boil up out of the crater, rise to 20,000 feet in a matter of seconds, and the entire sky was black with ash to the east as far as the eye could see.

INTRODUCTION

Mt. St. Helens is one of five volcanoes that dot the crest of the Cascade Range in Washington (Fig. 1). It is a geologically young volcano, dating back about 40,000 years, and has always been noted for its smooth, symmetrical cone (Figs. 2, 3). Its last eruption occurred in 1857. Minor steam eruptions were reported in 1898, 1903, and 1921.

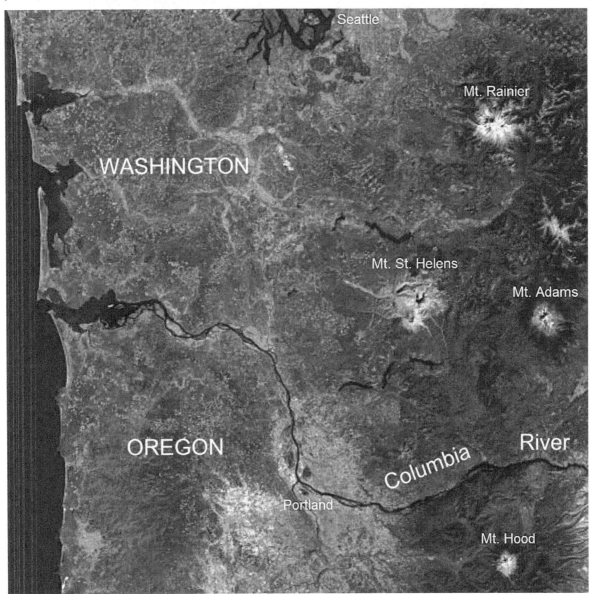

Figure 1. Location of Mt. St. Helens.

Figure 2. Pre-explosion volcanic cone of Mt. St. Helens. Spirit Lake in the foreground.

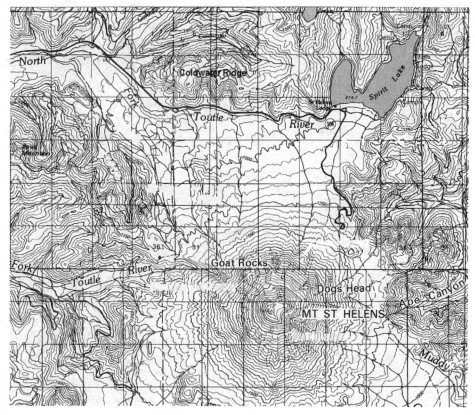

Figure 3. USGS topographic map of Mt. St. Helens prior to the 1980 eruption.

PRE–1980 ERUPTIONS OF MT. ST. HELENS

The first inkling that something might be brewing at Mt. St. Helens came on March 20, 1980 when a 4.2 magnitude earthquake occurred, followed by a swarm of earthquakes. Twenty four earthquakes of magnitude 4.0 or higher occurred in an 8-hour period on March 25. After March 25, earthquakes continued at a rate of 30 quakes of magnitude 3.0 or higher per day and 6 quakes of 4.0 or greater, the largest a magnitude 5.1 quake. In two days, 174 quakes of magnitude 2.6 or greater were recorded on seismographs.

At 12:36 p.m. on March 27, an explosive eruption of steam and ash occurred, sending an ash column about 7,000 feet into the air, mantling the flanks of the mountain with dark ash (Figs. 4, 5), and excavating a new crater about 250 feet across (Fig. 6). A large, down–dropped block between two fractures appeared at the summit of the mountain (Fig. 7).

Figure 4. Dark ash on Mt. St. Helens from the March 27 ash eruption. (Photo by Austin Post)

Figure 5. Ash on Mt. St. Helens from the March 27 ash eruption. (Photo by Austin Post)

Figure 6. New crater surrounded by ash after the March 27 ash eruption.
(Photo by Austin Post)

Figure 7. Down–dropped block along fractures at the summit of Mt. St. Helens. Dark ash surrounds a newly developed crater near the top of the photo. (Photo by Austin Post)

By March 29, a second crater had appeared next to the March 27 crater (Figs. 8, 9). Blue flames were visible from both craters, probably created by burning gases. On March 30, 93 eruptive outbursts were reported.

Figure 8. Twin craters in the down-dropped block bounded by fractures at the St. Helens summit. The bulge on the north flank of the cone is at the lower right. (Photo by Austin Post)

Figure 9. Ash–covered summit with dual craters and down-dropped summit block after the March 27 ash eruption. (Photo by Austin Post)

Strong harmonic tremors were detected on April 1, followed by explosive steam and ash eruptions (Figs 10-14). Harmonic tremor is a continuous quaking caused by injection of liquid magma into rocks below the surface. It is the most reliable indicator of an impending eruption.

Five earthquakes of magnitude 4 or greater occurred per day in early April. Explosive steam and ash continued on April 2 (Figs. 15-19).

Figure 10. Strong ash eruption April 1, 1980.

Figure 11. Ash eruption of Mt St. Helens, April 1, 1980.

Figure 12. Ash eruption, April 1, 1980.

Figure 13. Ash eruption and pyroclastic flow, April 1, 1980.

Figure 14. Closeup of pyroclastic flow rushing down the north flank of St. Helens, April 1, 1980.

Figure 15. Ash eruption, April 2, 1980.

Figure 16. Ash eruption, April 2, 1980

Figure 17. Ash eruption, April 2, 1980.

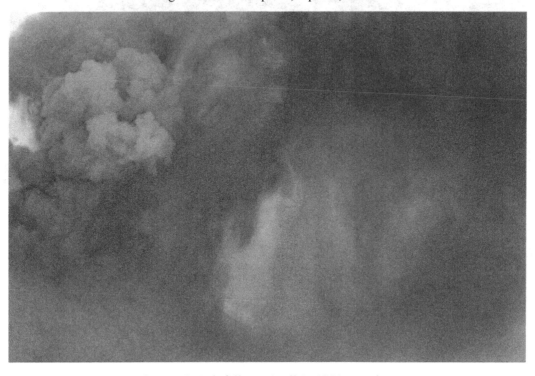

Figure 18. Ash fallout, April 2, 1980 eruption.

Figure 19. Ash covering the summit from April 2, 1980 eruption.

By April 7, the two summit craters had coalesced into a single large crater about 1,600 feet long by 1,000 feet wide and 500–800 feet deep. Earthquakes of 3.0 or greater continued at a rate of more than 30 per day. On April 10, an eruption occurred, mantling the cone with dark ash (Figs. 20-22).

Figure 20. Beginning of ash eruption on April 10. (Photo by AustinPost)

Figure 21. Ash eruption, April 10. (Photo by Austin Post)

Figure 22. Ash covered cone from April 10 ash eruption. (Photo by Austin Post)

Earthquakes of magnitude 3.2 or greater increased during April and May with eight per day the week before May 18. About 10,000 earthquakes were recorded prior to the May 18 event, mostly concentrated less than 1.6 miles directly below the cone (Fig. 23).

The down–dropped block in the summit cone continued to subside through April. By April 25, subsidence the summit was very distinct and the dual vents had coalesced into one deep crater (Fig. 24).

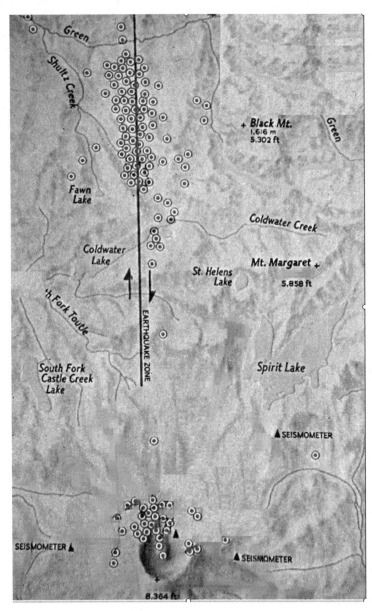

Figure 23. Earthquake epicenters (red dots) near Mt. St. Helens prior to the blast. (USGS)

Figure 24. Down–dropped block at the summit cone and deepening crater April 25.
(Photo by Austin Post)

Bulging of the north flank of the cone.

The north flank of St. Helens had begun to bulge outward in early April, and by the last week of April, an area 1.5 miles in diameter on the north flank had been displaced outward by at least 270 feet. During April and early May, the bulge grew outward 5 to 8 feet per day. By mid–May, the bulge extended more than 400 feet outward to north (Fig. 25). As the bulge moved outward to the north, the summit cone subsided, forming a down-dropped block. Goat Rocks, a rock knob low on the north flank, bowed outward 350 feet. By May 18, the north flank had bulged outward nearly 500 feet. Almost all of the motion was horizontal with very little upward component. The cause of the bulging was thought to be intrusion of magma into the cone

Figure 25. Bulging north flank of Mt. St. Helens. (Photo by Austin Post)

Figure 26. Ash–covered cone May 4. Mt. Adams in the background. (Photo by AustinPost)

Ash eruptions resumed May 7, the bulge continued to grow, 20–40 earthquakes per day occurred, including 5–10 of magnitude 4.0 or greater, and two periods of harmonic tremor occurred. Ash eruptions stopped on May 16, and the mountain was quiet until the May 18 blast (Figs. 26, 27).

THE EXPLOSION OF MT. ST. HELENS

On Sunday May 18, I scheduled a flight to St. Helens with Wilbur Johnson a, pilot and geologist. We planned to leave from Seattle about 7:00 a.m., but Wilbur liked sunrises, so we decided to leave at 5:00 a.m. instead. Our objective was to observe and photograph the down–dropped block and vent at the summit and the outward bulge of the north flank of the mountain. We arrived at the mountain about 5:30 and flew tight circles a few hundred feet over the summit and north flank of the mountain for more than two hours. On all other previous occasions when we had flown around the

24

volcano for several hours, it would eventually show some kind of eruptive activity. But on this day, the mountain was unusually quiet (Fig. 27), so we left to fly back to Seattle. About 20 minutes into the return flight, the plane suddenly bounced unexpectedly, and we joked about the mountain blowing up, not knowing that was indeed what had happened. Had we still been flying just above the summit, we would most likely have been blown out of the air, so Wilbur's love of sunrises may well have saved our lives.

When we landed in Seattle and learned what had happened, we flew back to St. Helens and photographed the eruption the rest of the day.

Figure 27. Quiet St. Helens volcano about half an hour prior to the explosion.

The lateral blast

At 8:32 a.m., 1980, a magnitude 5.1 earthquake triggered a massive landslide off the north flank of the volcano. The landslide travelled at 110 to 155 miles per hour and moved across the west arm of Spirit Lake. Part of the landslide overflowed 1,150–foot–high Coldwater ridge about 6 miles north of the mountain and spilled into the valley to the north. The rest of the slide flowed 13 miles down the Toutle River North Fork, filling the valley up to 600 feet deep with rock debris.

The collapse of the bulge generated a massive landslide that suddenly reduced the pressure on the molten rock beneath it, setting off an immense lateral explosion that quickly accelerated to more than 600 mph, perhaps briefly becoming supersonic.

The details of the May 18 explosion are accurately known from eye witness accounts and photos as the eruption unfolded. As I was leaving St. Helens less than half an hour before the explosion, Keith Stoffel, and his wife Dorothy were arriving there. As they were flying over the mountain, they witnessed the entire north flank of the volcanic cone collapse northward in a gigantic landslide A small ash eruption rose immediately from the base of the scarp made by the landslide, followed closely by the immense lateral blast.

A remarkable set of photos of the eruption, taken by Keith Ronholm from a ridge about 10 miles north of St. Helens, shows in great detail the sequence of events. A sobering thought for me is that if the explosion had occurred half an hour earlier or if we hadn't left the mountain when we did, we would have been in the midst of the blast.

The eruption sequence from the beginning of failure of the north flank of the volcano to the subsequent explosion has been made into a video by the U. S. Geological Survey (USGS). Clips from that video are shown in figures 28–31. A distant view of the explosion is shown in figure 32.

Figure 28. Beginning of the collapse of the bulge on the north flank of St. Helens. (USGS)

Figure 29. Massive failure of the bulge on the north flank of St. Helens. (USGS)

Figure 30. Beginning of the explosion. The dark plume is from the vent area and the gray plume is coming up from beneath the landslide. (USGS)

Figure 31. The explosion. The dark ash plume is coming from the vent area and the gray plume is coming up through the landslide. (USGS)

Figure 32. May 18 St. Helens blast, about 80,000 feet high and 40 miles wide.
(Photo by R. Kolberg)

A detailed description of the sequence of events has been reconstructed by Christiansen and Peterson (1981).

> *"the earthquake caused avalanching from the walls of the crater and only a few seconds later, triggered a sudden instability if the north flank. The entire north flank was described as "quivering" and appeared to almost liquefy. The slope failed along a surface intersecting the northern of the two high points on the north flank, near the east-west fracture separating the active bulge from the crater block. As the north flank began to slide away from this surface, a small, dark, ash-rich eruption plume rose directly from the base of the scarp and another from the summit crater rose to height of about 20m. As virtually the entire upper north flank slid off the cone and became a massive debris avalanche, a blast broke through the remainder of the flank, spewed ash and debris over a sector north of volcano, overtook the massive avalanche, and devastated an area nearly 30 km from west to east and more than 20 km northward from the former summit of the volcano. In an upper zone extending nearly 10 km from the summit, much of what had been densely forested, virtually no trees remained. Beyond, nearly to the limit of the blast, all standing trees were blown to the ground and at the blast's outer limit the trees were left standing but thoroughly seared. The devastated area of 600 km2 was blanketed by a deposit of hot debris carried by the blast.*

The sole of the debris avalanche was nearly at the base of the steep volcanic cone on the north side; the avalanche moved down the lower gradients of the volcano's outer flank and was nearly blocked by a ridge 8 km to the north. Part of the avalanche rounded the east end of that ridge and displaced the water from Spirit Lake, raising the bed in its southern part by more than 60 m. The bulk of the avalanche, however, turned westward down the valley of the North Fork Toutle River to form a craggy and hummocky deposit, part of which crossed the ridge to the north, but most of which flowed as far as 23 km down the North Toutle. The total volume of the avalanche in place is about 2.8 km³, and its length makes it one of the largest on record.

Water incorporated by the avalanche from North Fork Toutle River and possibly from Spirit Lake combined with melting blocks of ice from the torn-out glaciers of the volcano's north flank and melting snow and ice from the remaining slopes to produce mudflows that later in the day coursed across the avalanche and down the North Fork Toutle River, sweeping up thousands of logs from timbering operations in the valley and destroying most bridges across the river. The mudflows continued downstream, depositing sediment in the Cowlitz River channel and also obstructing the deep-water navigation channel of the Columba River."

The effects of the explosive eruption can be seen in several distinctive zones, mostly north of the mountain (Figs. 33, 34).

Figure 33. Mt. St. Helens crater and blast zone. (NASA)

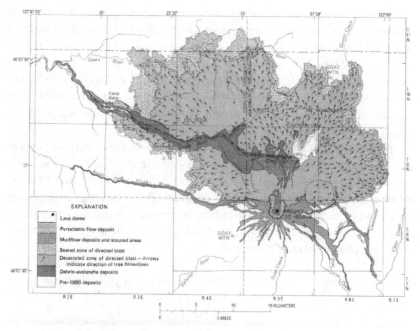

Figure 34. Map of zones of the May 18 eruption. (USGS)

The massive landslide that triggered the explosion was quickly overrun by the hot blast material, about 680°F, traveling at close to supersonic velocity. When the blast hit Spirit Lake, the water was displaced by the landslide and a 600-foot-high wave crashed into the ridge north of the lake. As the water sloshed back into the lake basin, thousands of trees were pulled into the lake where they remain today (Fig. 35).

The lateral blast totally devastated an area approximately 25 miles wide about 20 miles north of the volcano (Figs. 36-40. In the area a few miles north of the mountain, only splintered tree stumps remained where forests had been growing (Figs. 41-42), and the area was mantled with ash and landslide debris. Farther north, whole forests were blown down with trees draped like spaghetti over the hills. Still farther north, trees were seared by the blast and were killed, but remained upright. The near–supersonic lateral blast can be divided into three distinct zones:

1. Inner blast zone, averaging about 8 miles northeast and northwest of the volcano. Total devastation occurred within this zone (Figs. 43-46), only splintered tree stumps were left where living forest had previously existed and the area was covered with ash.

2. Outer blast zone, extending 19 miles NE and NW of the volcano in which whole forests were uprooted and blown down with their trunks aligned in great parallel swaths (Figs. 47-59).

3. Outermost dead–tree zone in which trees were seared by hot gases but remained standing (Figs. 60-61).

30

Figure 35. Spirit Lake, half filled with logs from the surrounding hills. (Photo by Austin Post)

Figure 36. Mt. St. Helens crater and the May 18 blast zone (bare ground).

Figure 37. Mt. St. Helens crater (foreground) and the May 18 blast zone (bare ground). Looking north, Mt. Rainier in the background.

Figure 38. St. Helens crater and blast area, looking eastward. Mt. Adams in the distance.

Figure 39. St. Helens crater and blast zone. (NASA image)

Figure 40. Devastated blast zone (bare ground), blown–down forest (brown), and dead-tree boundary with living forest (green).

Figure 41. Forested area north of St Helens prior to May 18, 1980. (Photo by Austin Post)

Figure 42. Same area devastated by the May 18 blast. (Photo by Austin Post)

Figure 43. Blast-devastated zone north of St. Helens. Spirit Lake in the center of the photo is half filled with logs (right of the open water). (Photo by Austin Post)

Figure 44. Blast area looking north from the crater. Log-filled Spirit Lake

Figure 45. Blast area looking north from the crater. Log-filled Spirit Lake in upper left.

Figure 46. Blast area looking north from the crater.

Figure 47. Splintered tree stumps from forest destroyed by the blast.

Figure 48. Splintered stumps of trees were blown away by the blast.

Figure 49. Splintered tree stumps from forest destroyed by the blast.

Figure 50. Splintered stump.

Figure 51. Splintered stump of what had been a tree before the blast.

Figure 52. Splintered tree stumps mantled with pyroclastic deposits from the May 18 blast.

Figure 53. Uprooted, blown–down trees mantled with pyroclastic deposits.

Figure 54. Trees blown down by the blast and draped over the topography like spaghetti.

Figure 55. Swirling pattern of trees blown down by the blast.

Figure 56. Trees along the valley side blown down by the lateral blast. Lateral levee of the landslide/lahar along the side of the valley.

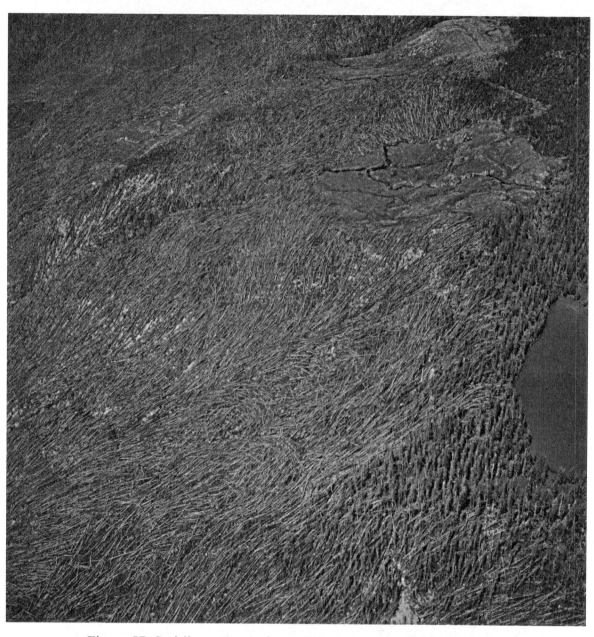

Figure 57. Swirling pattern of trees blown down by the lateral blast.
Note the sharpness of the boundary between the blown-down and standing trees.

Figure 58. Trees blown down by the lateral blast, Coldwater Ridge.

Figure 59. Tree blown down on the side of a ridge facing St. Helens. Note the dead trees still standing on the back side of the ridge. These trees were killed by the heat of the blast but were not blown over because they were protected by the ridge crest.

Figure 60. Forest blown down by the St Helens lateral blast. Note sharpness of the boundary between blow-downs and upright seared trees in the upper left, and the sharpness of the boundary between the seared trees and living trees in the upper left corner.

Figure 61. Sharp boundary between seared upright trees and living trees.

The ash plume

Immediately following the landslide off the north flank of the volcano and the massive lateral blast, a violent column of ash rose vertically from the crater to heights of about 80,000 feet (Figs. 62-66). An immense wall of black ash extended from the ground to 20,000 feet, driven eastward by high–altitude winds (Figs. 67-70). The ash reached Yakima, 90 miles to the east, by 9:45 a.m. (Figs. 71-73), depositing four to five inches of ash. The ash cloud extended to Spokane by 11:45 a.m., depositing half an inch of ash and plunging the area into darkness, reducing visibility to 10 feet. Ash fell in Yellowstone National Park at 10:15 p.m. and in Denver the next day.

Figure 62. Ash plume rising from the crater May 18, 1980.

Figure 63. St. Helens ash plume, May 18, 1980. (Photo by Austin Post)

Figure 64. Ash rising violently from the crater May 18, 1980.

Figure 65. Violent ash eruption, May 18, 1980.

Figure 66. May 18, 1980 eruption. Wall of ash extending eastward from the crater.

Figure 67. Black wall of ash east of Mt. St. Helens, May 18, 1980.

Figure 68. Black wall of ash east of the crater.

Figure 69. Ash plume and black wall of ash to the east, May 18, 1980.

Figure 70. Black ash plume, May 18, 1980.

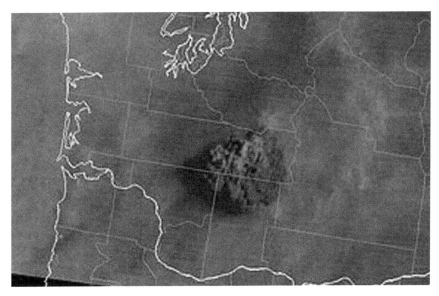

Figure 71. Satellite view of May 18 explosion of Mt. St. Helens. (NASA)

Figure 72. Satellite view of Mt. St. Helens ash plume. (NASA)

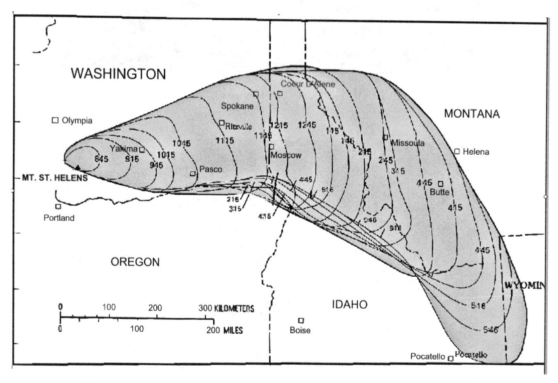

Figure 73. Chronology of expansion of the St. Helens ash plume.

During the first nine hours of the eruption, about 540,000,000 tons of ash were deposited over more than 22,000 square miles. Ash was deposited in 11 states during the following two weeks (Fig. 74). One day, we noticed an unusual accumulation of dust on car windows in Bellingham. Under the microscope it proved to be St. Helens ash (Fig. 75), but none of the ash plumes from St. Helens reached this far north, so we concluded that it was ash that had drifted eastward all the way around the globe before falling out in Bellingham.

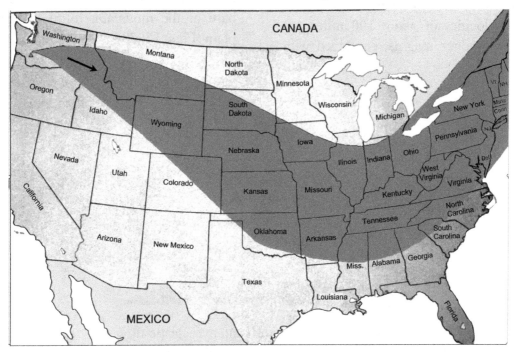

Figure 74. Ash fallout plume from St. Helens.

Figure 75. Microscopic view of pumice particles in ash.

Pyroclastic flows

Pyroclastic flows (Greek pyros=fire; clastos=broken) are masses of very hot gasses, molten rock, pumice, volcanic glass, and rock fragments that are thrown into the air by explosive volcanic eruptions then rush down the sides of volcanoes (Figs. 76, 78) at velocities of about 100 miles per hour. The molten magma, combined with hot gasses and temperatures over 1,000° F, makes these eruptions extremely fluid. They are blown upward out of the volcanic vent and are pulled rapidly downslope by gravity (Fig. 78). The pyroclastic flows erupted from St. Helens came to rest at the base of the mountain, forming a pumice plain (Figs. 77, 79).

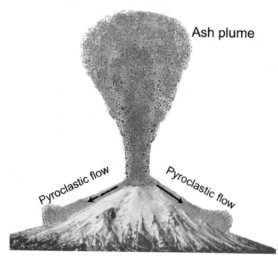

Figure 76. Eruption of ash and pyroclastic flows.

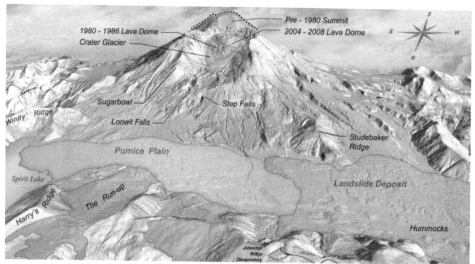

Figure 77. St. Helens crater, pyroclastic pumice deposits and landslide deposits. (USGS diagram)

Figure 78. Pyroclastic flow erupted from the crater
flowing down the north slope of the volcano. (USGS photo)

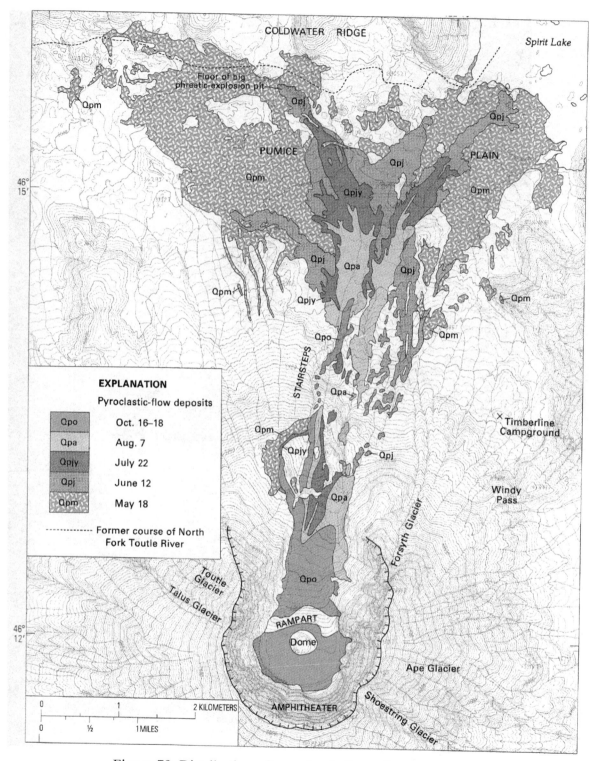

Figure 79. Distribution of pyroclastic flows from St. Helens between May 18 and October, 18, 1980. (USGS)

Pyroclastic flows began erupting about four hours after St. Helens exploded and continued for about five hours. Multiple pyroclastic flows were emitted from the open end of the crater and flowed down the north slope of the cone for distances up to five miles. They ranged in thickness from a few feet to 30 feet and were emplaced at temperatures of about 1,300° F. They were still about 575 to 800° F two weeks after they were erupted (Figs. 86-88).

Additional pyroclastic flows were subsequently erupted during five major eruptive episodes that followed the May 18 explosion: May 25, June 12, July 22, August 7, and October 16-18. The pyroclastic flows formed long, digitate lobes of ash, pumice, and blocks of lava (Figs. 82-85). The lobes spread out into fan-like patterns when they reached the pumice plain north of the volcanic cone (Figs. 80-81). Fumaroles and steam explosion pits formed where the flows buried lakes or streams, flashing the water into steam that erupted up through the pyroclastic flows. The surfaces of many of the pyroclastic flows are pitted with depressions from these localized steam eruptions (Fig. 89). Most pits are circular with diameters of 15 to 300 feet and 3 to 60 feet deep. The largest pit was 2,300 feet long, 1,000 feet wide, and 125 feet deep.

Figure 80. Pyroclastic pumice plain north of the St. Helens crater.

Figure 81. Terminus of a long, pyroclastic pumice flow on the floor of the Toutle valley.

Figure 82. Lobe of pyroclastic pumice flows in the upper Toutle Valley.

Figure 83. Digitate lobes along the margin of a pyroclastic pumice flow in the Toutle Valley

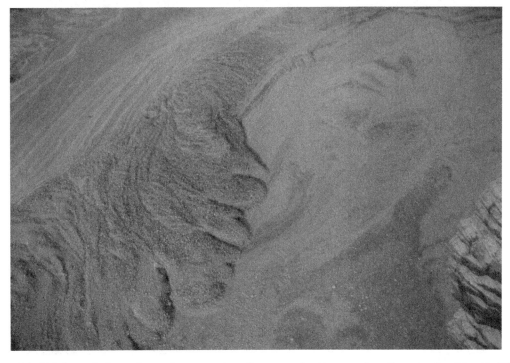

Figure 84. Digitate lobes along the margin of a pyroclastic pumice flow in the Toutle Valley

Figure 85. Digitate margins of overlapping pyroclastic pumice flows, upper Toutle valley. (USGS photo)

Figure 86. St. Helens crater with pumice stretching out from the base.

Figure 87. Pyroclastic pumice flow about a month old.

Figure 88. Close-up of pumice clasts making up the pyroclastic pumice flow in Fig. 86. The surface was cool, but less than a foot below the surface, the pumice was too hot to touch a month after deposition of the pumice.

Figure 89. Surface of a pyroclastic flow pock-marked by steam explosion pits, south end of Spirit Lake. Note the logs floating in the lake (left side of the photo) (Photo by Austin Post)

Landslides and lahars (volcanic debris flows)

The massive landslide that resulted from failure of the north flank of the volcano was literally blown apart by the explosion that immediately followed. This landslide has been described as world's largest. Heat from the exploding magma melted most of the glaciers on the north side of the mountain and that water helped generate a huge volcanic mudflow (lahar) that sped down the slopes at close to 100 mph before slowing on the lower, flatter part of the Toutle valley. The landslide/lahar filled the valley of the North Toutle River to a maximum depth of 600 feet, averaging 150 feet, and extended 14 miles downvalley (Figs. 90-128). The total volume of the deposit is about 3.3 billion cubic yards, enough to fill about a billion dump trucks.

The landslide/lahar destroyed the logging facility at Camp Baker (Fig. 128), picking up many logs that were carried downstream. Twenty seven bridges and about 200 homes were destroyed. When the lahar entered the Columbia River, it reduced its depth by 25 feet for four miles, temporarily closing the channel to large ships.

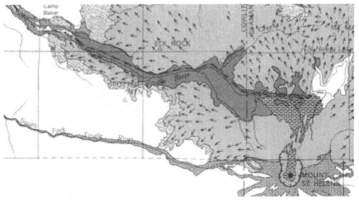

Figure 90. Map of blast zone and huge landslide/lahar (dark color) in the valley of the North Fork of the Toutle River. Arrows indicate direction of the lateral blast. (USGS)

Figure 91. Mt. St. Helens landslide/lahar filling the valley of the Toutle North Fork. (Photo by Austin Post)

Figure 92. Massive landslide/lahar filling the Toutle valley downvalley from Mt. St. Helens. (Photo by Austin Post)

Figure 93. Massive landslide/lahar filling the Toutle valley downvalley from Mt. St. Helens.

Figure 94. Massive landslide/lahar filling the upper Toutle Valley.

Figure 95. Massive landslide/lahar filling the upper Toutle Valley.

Figure 96. Large, irregular hummocks on the surface of the landslide/lahar.

Figure 97. Large, irregular hummocks on the surface of the landslide/lahar.

Figure 98. Large, irregular hummocks and depressions on the surface of the landslide/lahar.

Figure 99. Landslide/lahar filling the Toutle valley. The bare area along the valley sides were deforested by the blast.

Figure 100. Landslide/lahar filling the Toutle valley.

Figure 101. Landslide/lahar filling the Toutle valley.

Figure 102. Thickness of the landslide/lahar exposed by incision of the upper Toutle River.

The landslide/lahar consists of blocks and debris from the landslide mingled with pumice from the volcanic explosion (Fig. 103). The landslide/lahar averages 125 feet thick, reaching 600 feet in its upper part. Its upper surface is very irregular, with huge blocks of rock standing well above the surface and many smaller mounds of rock debris.

Figure 103. Landslide/lahar deposit composed of slide debris and pumice.

Figure 104. Hummocky surface and small lakes on the landslide/lahar, upper Toutle Valley.

Figure 105. Large block from the flank of the volcano carried down the
Toutle Valley by the landslide/lahar.

Figure 106. Large block from the flank of the volcano on the surface of the landslide/lahar.

Figure 107. Blocks of volcanic debris on the surface of the landslide/lahar.

Figure 108. Large mound of landslide debris on the surface of the landslide/lahar.

Figure 109. Logs embedded in the landslide/lahar.

Figure 110. Splintered stumps and landslide/lahar deposits.

Figure 111. Lahar levee along the valley side in the blast zone.
Note the trees blown down by the blast.

Figure 112. Levee of the massive landslide/lahar along the valley side of the Toutle River.

Figure 113. Landslide deposits on a divide, upper Toutle Valley.

Figure 114. Landslide debris along in the upper Toutle drainage.

Figure 115. Small lake impounded by the landslide/lahar damming a side stream.

Figure 116. Landslide/lahar in a side stream within the blast zone

Figure 117. The lower, more fluid part of the lahar in the lower Toutle Valley.

Figure 118. Fluid mudflows filling in around hummocks on the surface of the lahar in the lower Toutle Valley.

Figure 119. Fluid mudflows filling in around hummocks on the surface of the lahar in the lower Toutle Valley.

A helicopter pilot flying rescue missions on May 18 took a series of remarkable photos of a bridge in the Toutle valley (Figs. 120-123). Figure 119 shows people standing on a bridge watching a steaming lahar pass under it. Then things turned ugly and the lahar began flowing over the bridge (Fig. 120) and sweeping it downstream (Fig. 121-122).

Figure 120. People standing on a bridge watching a steaming lahar passing beneath it.

Figure 121. The lahar begins to flow over the bridge.

Figure 122. The lahar sweeps away the bridge.

Figure 123. The bridge being swept downstream by the lahar.

Figure 124. Final resting place of the bridge.

As a series of lahars pulsed down the Toutle valley, they incorporated river water and became more fluid and caused flooding in the lower parts of the valley. Because of the large sediment load they carried, the river bottom silted up rapidly (Fig. 125-128), filling the channel and floodplain with silt and sand.

Figure 125. Mudflow filling the lower Toutle valley.

Figure 126. Mudflow in the lower Toutle valley.

Figure 127. Mudflow trim lines on trees showing the height of lahars.

Figure 128. Mudflow trim line on trees in the lower Toutle Valley.

Figure 129. Logs at Camp Baker were swept away by a lahar.

Figure 130. Toutle River choked with logs swept away by floods.

Figure 131. Soupy lahar flooding the lower Toutle valley

FATE OF MT. ST. HELENS GLACIERS

Prior to the May 18 explosion of Mt. St. Helens, its sides were mantled with a dozen glaciers, fed by snowfall high on the mountain. All of these glaciers were beheaded by the May 18 explosion when the upper 1,314 feet of the mountain was blown off and the source areas of the glaciers were destroyed. Figures 132 and 133 show the glaciers before the May 18 explosion, and Figure 134 shows the post-eruption remnants. Figure 135 shows the Shoestring glacier prior to the eruption,

Figure 132. Glaciers on Mt. St. Helens before the May 18 explosion. (Photo by Austin Post)

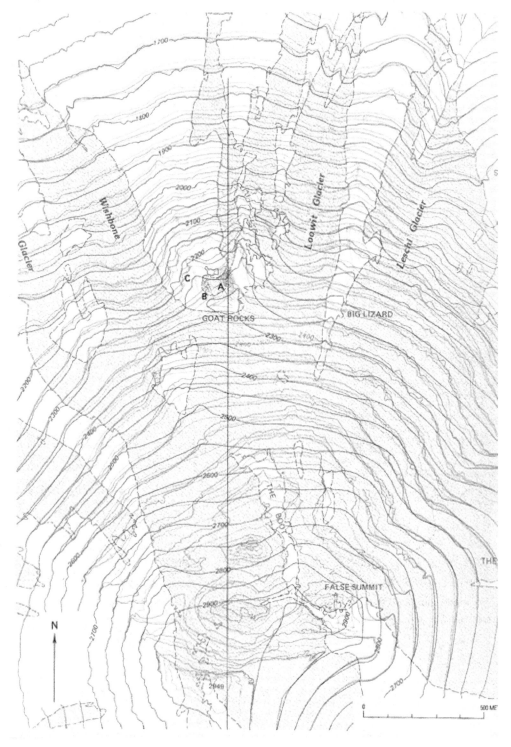

Figure 133. Glaciers on Mt. St. Helens before the 1980 eruption. (USGS)

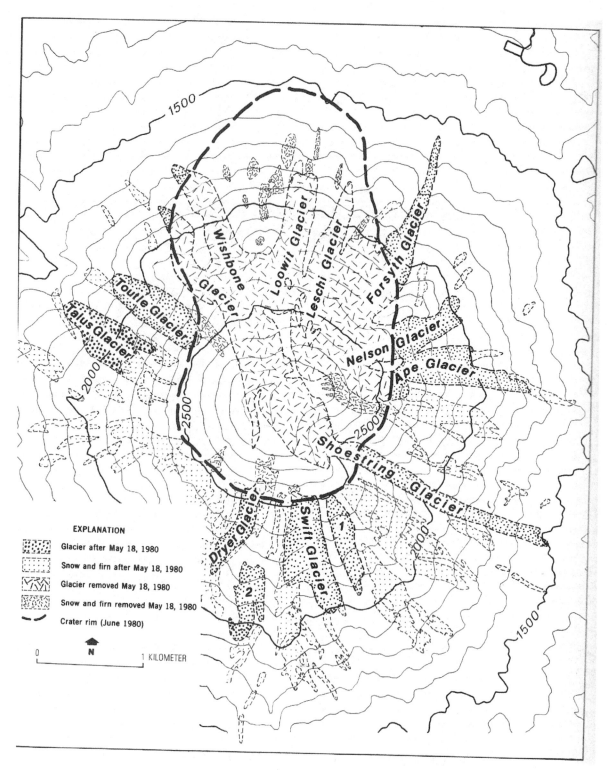

Figure 134. Destruction of glaciers on Mt. St. Helens as a result of the 1980 explosion. (USGS)

Figure 135. Shoestring glacier prior to the May 18 eruption. (Photo by Austin Post).

A comparison of Mt. St. Helens glaciers before and after the May 18 explosion is shown on figures 133 and 134. The Leschi, Loowit, and Wishbone glaciers were entirely destroyed by the May 18 blast and the Forsyth, Nelson, Ape, and Shoestring glaciers were beheaded.

Figure 136. Beheaded remnants of the Shoestring glacier following the May 18 blast.
(Photo by Austin Post)

Dome Building, Later Eruptions

On May 25, 1980, a sudden increase in earthquake activity preceded an eruption that sent ash about 50,000 feet into the atmosphere and pyroclastic flows went out the northern breach of the crater, covering earlier pyroclastic flows and landslide and lahar debris. Large areas in western Washington and Oregon were lightly dusted with ash.

Five explosive eruptions of ash occurred between May and October 1980. Following ash eruptions in June, August, and October, very viscous, pasty lava pushed up extrusive domes on the crater floor (Figs. 137-139). On June 12, 1980, several ash eruptions rose to about 50,000 feet, coating the Portland area with ash.

Following the ash eruption, a 200–foot high, 1,200–foot wide, lava dome was pushed up through the crater floor. The lava dome consisted of stiff, pasty lava, covered with a crust of earlier–cooled lava (Fig. 137).

After more than a month of quiescence, renewed activity began on July 22, following increased earthquake activity, expansion of the summit area, and change in emission rates of sulfur dioxide and carbon dioxide. An ash eruption rose more than 50,000 feet (Fig. 140, blowing away the dome in the crater.

On August 7, 1980, following increased seismic activity and gas emission, an ash eruption rose more than 40,000 above the crater. Pyroclastic flows erupted through the northern breach in the crater wall. An extrusive dome about 200 feet high and about 400 feet in diameter was rebuilt on the crater floor.

Figure 137. Pasty, molten lave pushing up into an extrusive dome.

Figure 138. Extrusive dome on the crater floor. Spirit Lake in the foreground. (USGS photo)

Figure 139. Extrusive dome in St. Helens crater. (USGS photo)

Figure 140. July 22, 1980 ash eruption. (USGS photo)

After two months of quiescence, renewed eruptions occurred from October 16 to 18, 1980. Ash was erupted to more than 50,000 feet, accompanied by small pyroclastic flows, and the August dome was blown away. A new dome began to form within 30 minutes after the final ash eruption on October 18 and within a few days, grew to about 130 feet high and 900 feet wide. Figure 141 shows the dome October 14, 1981. By early January, 1981, the dome had developed into two lobes, the larger of which was about 300 feet high and 650 feet in diameter. By 1987, the dome had grown to more than 800 feet high and 3,000 feet wide.

Figure 141. Lava dome in St Helens crater Oct. 14, 1981. (Photo by Austin Post)

Figure 142. Lava dome building in St. Helens crater, May 24, 1982. (Photo by Austin Post)

From the October 1980 eruptions to 1986, a new lava dome 876 feet above the crater floor was built during 17 eruptive episodes (Fig. 142). In 1983–84, molten rock was injected into the interior of the dome, pushing it eastward 250 feet.

About 30 brief, intense, seismic outbursts occurred between 1989 and 1991. Several small ash eruptions from a new vent on the north side of the dome hurled three–foot boulders half a mile northward from the dome. Small pyroclastic flows were also generated.

Following a decade of quiescence, a new period of eruptive activity began in 2004 and ended in 2008. From September 23 to 29, 2004, swarms of earthquakes of 2.5 or greater began to occur at rates of four per minute. On October 1, 2004 an ash eruption (Fig. 143) rose to 10,000 feet.

A glacier developed on the floor of the crater during the decade-long quiet period leading up to 2004. In 2004, a new dome pushed up through the glacier, piercing the glacier and pushing it laterally and causing many fractures in it (Figs. 143-144).

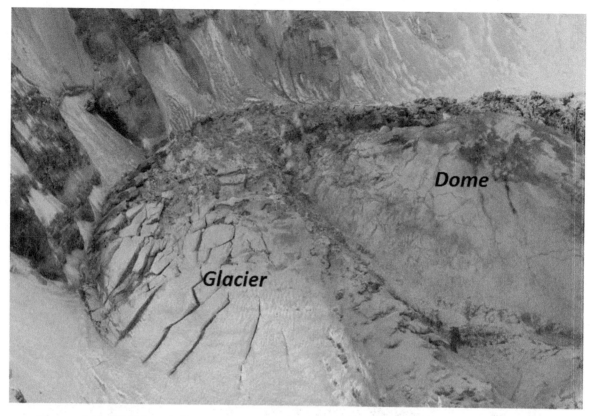

Figure 143. Uplift and fracturing of a glacier against south crater wall as a dome pushes upward, November 29, 2004. (Photo by Jim Vallance and Matt Logan, USGS)

Figure 144. Fracturing of a lava dome pushing up on the crater floor, December, 28, 2004. (Photo by John Pallister, USGS)

Figure 145. Ash eruption October 5, 2004. (Photo by Steve Chilling, USGS)

About October 11, new lava dome began to form and continued to grow until January, 2008. The new dome consisted of seven spines with polished sides showing that they had been pushed up as solids. Figure 146 shows the configuration and timing of the emplacement.

Part of the new dome was a feature called a "whaleback," so named because of its resemblance to the back of a whale (Figs. 147, 148), consisting of a hot mass of solid lava being extruded by pressure of the molten rock beneath it.

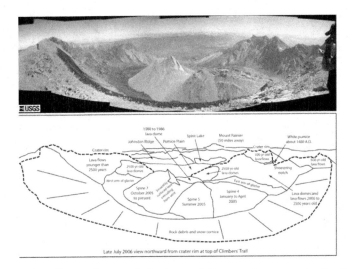

Figure 146. Configuration and timing of the emplacement of spines in the lava dome on the crater floor. (USGS)

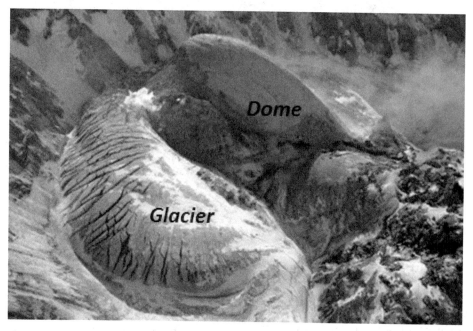

Figure 147. Extrusive dome of viscous, pasty lava pushing up through the glacier on the crater floor (fractured area in left center) February 22, 2005. Whaleback–shaped lava spine at top center. (Photo by Steve Shilling USGS)

Figure 148. 'Whaleback' feature on the extrusive dome on the crater floor, February 22, 2005. (Photo by USGS)

By February 1, 2005, the new extrusive lava dome on the crater floor was about 2,000 feet above the surface of the new glacier on the crater floor. The diameter of the new dome was about 1,700 feet and the whaleback feature was 1,550 feet long and 500 feet wide. On March 8, 2005, a brief steam and ash eruption rose to 36,000 feet.

As the new dome continued to grow, the whaleback feature gradually disintegrated. The end of the 'whaleback feature' broke off on July 2, 2005, sending a cloud of ash and dust several hundred feet into the air. Part of the dome rose at a rate of up to 6 feet per day. Growth of the dome ended in 2008. Figure 132 shows the dome in 2016.

Figure 149. Extrusive dome on the crater floor in 2016.

The May 18, 1980 explosion of Mt. St. Helens was the most awesome natural event to occur in North America in historic time. It was the most destructive volcanic event in US history. About 57 people were killed, and 200 houses, 47 bridges, and 185 miles of highway were destroyed. Four billion board feet of timber were destroyed.

Interstate 90 from Seattle to Spokane was closed for a week and a half, airports in eastern Washington were closed, and more than 1000 commercial flights were canceled. An estimated 2.4 million cubic yards of ash, roughly equivalent to more than two million dump trucks, were removed from airports and highways

REFERENCES

Brantley, S. R. and Myers, B., 2000, Mount St. Helens—From the 1980 Eruption to 2000: U.S. Geological Survey Fact Sheet, FS-036-00, 2 p.

Christianson, R.L. and Peterson, D.W., 1981, Chronology of the 1980 eruptive activity: *in* Lipman, P.W. and Mullineaux, D.R., eds., The 1980 eruptions of Mount St. Helens, Washington, U.S. Geological Survey Professional Paper 1250, p. 17-30.

Cummans, J., 1981, Mudflows resulting from the May 18, 1980, eruption of Mount St. Helens, Washington: U.S. Geological Survey Circular, 850-B, 16 p.

Foxworthy, B. L. and Hill, M., 1982, Volcanic eruptions of 1980 at Mount St. Helens: The First 100 Days. U.S. Geological Survey Professional Paper 1249, 125 p.

Hoblitt, R.P., Miller, C.D., and Vallance, J.W., 1981, Origin and stratigraphy of the deposit produced by the May 18 directed blast: *in* Lipman, P.W. and Mullineaux, D.R., eds., The 1980 eruptions of Mount St. Helens, Washington, U.S. Geological Survey Professional Paper 1250, p. 401-419.

Janda, R.J., Scott, K.M., Nolan, K.M., and Martinson, H.A., 1981, Lahar movement, effects, and deposits: *in* Lipman, P.W. and Mullineaux, D.R., eds., The 1980 eruptions of Mount St. Helens, Washington, U.S. Geological Survey Professional Paper 1250, 461-478.

Krimmel, R.M., 1981, Oblique aerial photography, March-October 1980: *in* Lipman, P.W. and Mullineaux, D.R., eds., The 1980 eruptions of Mount St. Helens, Washington, U.S. Geological Survey Professional Paper 1250, p. 31-51.

Lipman, P. W. and Mullineaux, D. R., eds., 1981, The 1980 eruptions of Mount St. Helens, Washington. U.S. Geological Survey Professional Paper, 1250, 844 p.

Moore, J.G. and Sisson, T.W., 1981, Deposits and effects of the May 18 pyroclastic surge: *in* Lipman, P.W. and Mullineaux, D.R., eds., The 1980 eruptions of Mount St. Helens, Washington, U.S. Geological Survey Professional Paper 1250, p. 421-438.

Mullineaux, D.R. and Crandell, D.R., 1981, The eruptive history of Mount St. Helens: *in* Lipman, P.W. and Mullineaux, D.R., eds., The 1980 eruptions of Mount St. Helens, Washington, U.S. Geological Survey Professional Paper 1250, p. 3-15.

Rowley, P.D., Kuntz, M.A., and Macloud, N.S., 1981, Pyroclastic flow deposits: *in* Lipman, P.W. and Mullineaux, D.R., eds., The 1980 eruptions of Mount St. Helens, Washington, U.S. Geological Survey Professional Paper 1250, p. 489-512.

Sherrod, D. R., Scott, W. E., and Stauffer, P. H., eds., 2008, A volcano rekindled: The renewed eruption of Mount St. Helens, 2004-2006: U.S. Geological Survey Professional Paper, 1750, 856 p.

Voight, B., 1981, Time scale for the first moments of the May 18, 1980 eruption: *in* Lipman, P.W. and Mullineaux, D.R., eds., The 1980 eruptions of Mount St. Helens, Washington, U.S. Geological Survey Professional Paper 1250, p. 69-86.

Voight, B., Glicken, H., Janda, R.J., and Douglass, P.M., 1981, Catastrophic rockslide avalanche of May 18: *in* Lipman, P.W. and Mullineaux, D.R., eds., The 1980 eruptions of Mount St. Helens, Washington, U.S. Geological Survey Professional Paper 1250, p. 347-377.

CPSIA information can be obtained
at www.ICGtesting.com
Printed in the USA
FSOW04n1500191116
27562FS

9 780692 649911